Vehicle Security Systems

Vehicle Security Systems:

Build your own alarm and protection systems

A L Brown

B⊞ NEWNES

Newnes
An imprint of Butterworth-Heinemann Ltd
Linacre House, Jordan Hill, Oxford OX2 8DP

℞ A member of the Reed Elsevier plc group

OXFORD LONDON BOSTON
NEW DELHI SINGAPORE SYDNEY
TOKYO TORONTO WELLINGTON

First published 1996

© Butterworth-Heinemann Ltd 1996

British Library Cataloguing in Publication Data
A catalogue record for this book is available from the British Library
ISBN 0 7506 2630 5

Library of Congress Cataloguing in Publication Data
A catalogue record for this book is available from the Library of Congress

Printed and bound in Great Britain by
Biddles Ltd, Guildford and King's Lynn

Contents

7 A multi-purpose garage/outhouse alarm system 58

8 Experimental circuits for remote control 71

Appendix: PCBs   79

Introduction

When designing an alarm system, the needs of the user are of prime importance. Almost as important is understanding the potential car thief. Adding the two together should result in an effective security system for combating car crime.

With these requirements in mind the design of the alarm is built around the objective as stated below.

The design objective

To design a self-activating security system which does not require the driver to do anything other than to enter or vacate the vehicle within certain time limits, to immobilize the car and sound the alarm if entered or tampered with by an unauthorized person or persons....

Although the circuits explained in 'Initial Defences' do not strictly comply with the objective they are all effective on their own and combined together they could form a cheaper alternative to a full alarm system.

Remote vs self-activating

'An alarm system must be switched on to be effective'.

This statement may seem obvious but when I think of how many times I have heard people say 'I didn't think it was worth putting the alarm on, after all I only stopped at the shop for a paper'.

Consider also a woman who has had her handbag stolen. If the handbag contained her car keys, etc., and should the car alarm in use be of the remote type which responds to the control by flashing the car lights and giving a few 'blips' on the siren, how long would it be before the thief located the car by walking around the local car parks using the remote control while looking and listening for a response? Unfortunately the remote control of the key-ring will also tell the car thief which type of alarm he will encounter, thus making it all the easier to steal the car.

If, however, the alarm is not of the remote type then there will be no way of determining to which car the keys belong. It would be a monumental task to check out all of the cars of the same make in any given location. Not understanding the system in use will also seriously reduce the chances of the car being stolen.

With a self-activating alarm system the car would always be protected unless the alarm were switched off for servicing work, etc.

For the constructor who requires a security system which attempts to cover most ways of stopping car theft and which should give peace of mind, I will describe several different pieces of equipment. These will allow the constructor to decide the level of security he requires depending on his individual needs and circumstances.

Starting at the most basic level there is a simple flashing warning beacon controlled by the ignition key. Following that are some immobilizers to safeguard your car. A few different ideas are given with the options for engine disablement left to the constructor.

Next on offer is a basic alarm which is very reliable and easy to use. This protects the car by sounding the alarm immediately if the boot or bonnet are tampered with or after a short delay should a door be opened. The warning beacon and an immobilizer are both incorporated into this design to increase its capabilities. This alarm is most suitable for 'soft top' or 'cabriolet' types.

For those who require further protection, another more sophisticated design is described. This version includes ultrasonic sensors. Other add-on devices are included to enhance the system, and again the constructor has control over the level of security needed.

With the construction and installation details provided, building the systems described should be within the reach of most constructors. A fault diagnosis section used together with the circuit diagrams and flow chart should help with any problems that may arise during testing or installation.

The alarm system described in this book should prove to be an effective ally in the fight against car crime. Once the alarm has decided the car is being tampered with it changes from being user-friendly and becomes a hard to beat unit determined to stop the car from being stolen.

Chapter 1

Initial defences

The warning beacon

The first line of defence in car security is a visible indicator showing there is some form of security fitted to the car. Using a flashing LED (light emitting diode) as a warning beacon can be very effective. Without any form of warning device the car thief will not know there are any security devices fitted and will be just as likely to break into your car as any other. If the beacon is seen then a softer option parked down the road may be stolen instead.

A brief circuit description

Figure 1.1 shows the circuit for a simple flashing beacon using the common 555 timer i.c. (integrated circuit) in astable mode. With the addition of a transistor (Q1) we can control the operation of the timer with the ignition key.

When the ignition is turned on Q1 is biased on by the resistor network formed by R5 and R6. Pin 4 is held low by the collector of Q1 keeping it below the 0.6 V threshold above which the timer will oscillate. Initially the output pin (3) is low and the LED is off. Discharge pin (7) is low keeping the timing capacitor from charging towards VCC (positive supply voltage) and pin 2 below the trigger voltage of $\frac{1}{3}$ of VCC.

If the ignition is turned off then the collector of Q1 releases pin 4

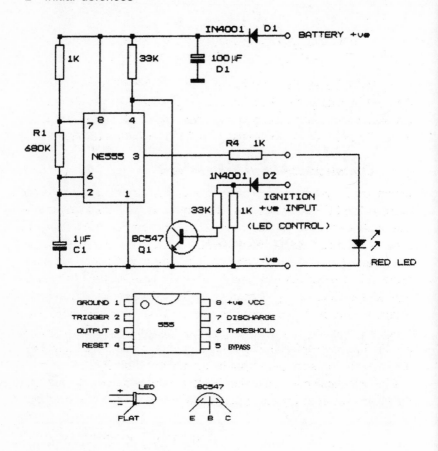

Figure 1.1 Flashing beacon LED

causing it to go high to VCC. With pin 4 above 0.6V and pin 2 below $\frac{1}{3}$ of VCC the timer triggers and pin 7 allows C1 to start charging. At this point pin 3 goes high and turns on the LED.

As the charge on C1 reaches $\frac{2}{3}$ of VCC the threshold detector switches the internal flip-flop causing pin 7 to go low discharging C1 and pin 3 top switches off the LED. When the charge on C1 falls to $\frac{1}{3}$ of VCC pin 2 triggers the timer and the sequence is restarted.

The frequency is determined by the formula:

$$f = 0.72 / C1 \times R3$$
$$(C1)\ 0.000001\ \text{farad} \times (R3)\ 680\,000\ \text{ohms} = 0.68$$
$$f = 0.72 / 0.68 = 1.05\ \text{Hz}$$

D1 and D2 ensure correct supply polarity. The resistor R6 reduces the input impedance to Q1 and discharges any capacitors on the ignition line which can cause a delay in the operation of the timer circuit. R4 limits the LED current to around 12 mA C2 decouples the supply.

Construction and installation

In Figure 1.2 there is a printed circuit board (PCB) layout and component overlay. The PCB can be copied either by hand or by photographing it onto a photosensitive circuit board. (See Appendix for PCBs drawn to scale.) Fit the components onto the PCB and connect the supply wires using the colour codes listed on the circuit diagram. Connect the black lead to a 12 V negative supply and the red to the positive. The LED should light up for about two seconds and then start flashing. By touching the yellow control lead to the positive line the flashing should stop. This simulates switching on and off the ignition when the unit is fitted into the car. Fit the completed unit into a small plastic box and use a cable tie to hold the wires to stop them from being pulled off the PCB.

Find a suitable position under the dashboard and make the box and the cables secure using cable ties. Make good, tight connections and

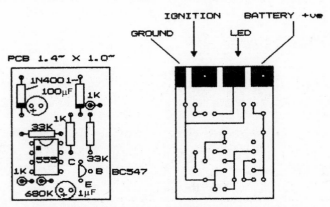

Figure 1.2 Flashing beacon PCB and component overlay

use PVC tape for insulation. Most cars have the connections you need behind the radio which can make things easy! Make sure the LED is visible from most positions outside of the car. Try and find a spare switch blank rather than drill the dashboard and use an LED grommet in the hole to finish off the job properly.

The LED should have started flashing immediately the red and black wires were connected and stopped as the ignition was switched on with the yellow wire connected.

Remember to keep the wiring neat and tidy and well away from moving parts such as pedals and heater controls.

The immobilizer

There are many ways to immobilize a car and the most popular are listed below. Any one, or a combination, of these can be effective against car theft.

1   Disconnection of the ignition positive supply.
2   Disconnection of the starter relay (solenoid).
3   Disconnection of the petrol pump (if electrical).
4   Disconnection of the diesel fuel valve.
5   Insertion of a resistor into the ignition negative lead.

Apart from using a toggle switch (which does not meet the requirements of the design philosophy) the simplest method of immobilizing a car is to use a 12 V double pole relay as shown in Figure 1.3.

The circuits automatically reset when the engine is stopped and restarting the engine is only possible if PB1 is pressed, thus completing the coil circuit. When the relay energizes contacts (A) bypass PB1 holding the relay energized until the ignition is again turned off. Contacts (B) can be used to control any of the above functions.

In the case of item No 5, a 33 ohm 7 W resistor is used to limit the current in the coil primary thereby reducing the voltage in the secondary and the likelihood of any sparks at the sparkplugs. An advantage of this method is that 'hot-wiring' the ignition will not work and providing the wiring is neat and the connections are well hidden then it may well

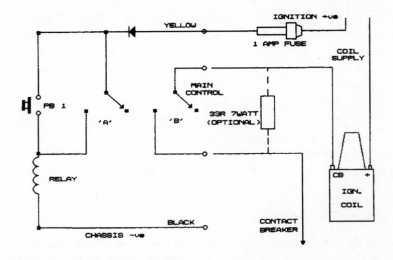

Figure 1.3 Basic pushbutton immobilizer (circuit)

baffle a potential car thief. Figure 1.4 shows a suitable printed circuit layout and component overlay. On expanding the system to cover more than one of the items listed, it would be necessary to use either a four pole relay or two double pole relays in parallel.

An improvement would be the addition of a waterproof siren fitted

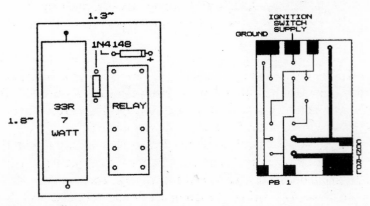

Figure 1.4 Basic pushbutton immobilizer, with optional 33R resistor (PCB and layout)

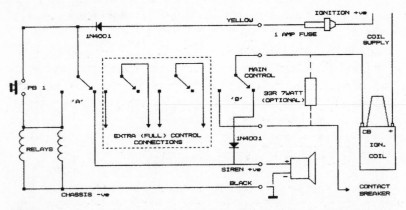

Figure 1.5 Pushbutton immobilizer with full control and siren (circuit)

under the bonnet to warn of attempted theft of the car. This should be connected as shown in Figure 1.5. When the ignition is switched on, power is applied to the siren via contacts (A) but as the push button (PB1) is pressed and the relay is energized the contacts (A) change over, switching off the siren and switching on the ignition supply to the coil as before. To make life even harder for the car thief an internal sounder can be fitted inside the car to create a deafening wall of sound

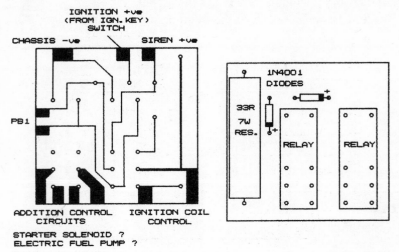

Figure 1.6 Pushbutton immobilizer with full control and siren (PCB and layout)

to discourage any unauthorized tampering with the vehicle ignition circuits.

Possible modifications are shown in Figure 1.6 and include both sounders. Fit the external siren under the bonnet taking care to keep the wiring neat and tidy and well away from moving parts. The internal sounder should be mounted in a relatively inaccessible position safe from attack by an intruder.

Electronic 'key' immobilizer

Although the push button would in practice be well hidden, it is always possible that it may be found, thus reducing the effectiveness of the immobilizer.

To overcome this problem the circuit in Figure 1.7 describes an 'electronic combination key' immobilizer using standard dim plugs as the 'key' and a matching panel socket as the 'lock'. The plug and socket are wired to match each other to form the correct combination switching on the transistors Q1 and Q2. If, however, the incorrect combination is applied, then Q3 will switch on, inhibiting the relay from operating and will therefore stop the car from being started. Q3 can also be used to trigger a separate alarm system using the output provided.

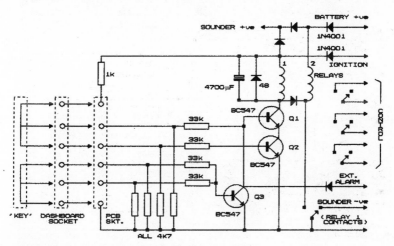

Figure 1.7   Electronic 'key' immobilizer (circuit)

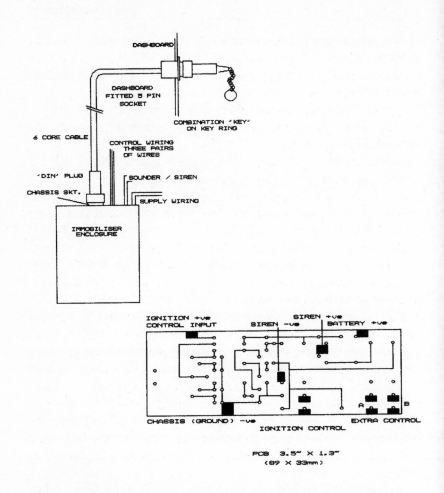

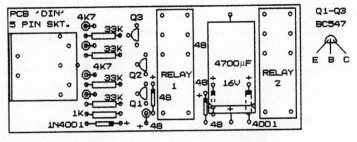

Figure 1.8 Electronic 'key' immobilizer (PCB and layout)

A 4700 μF capacitor wired across the relay coil will hold the relay energized for a couple of seconds if the car has stalled and needs to be cranked over again quickly, but only if the 'key' was not kept in the immobilizer socket.

Figure 1.8 illustrates a printed circuit board layout and a component overlay suitable to cover any combination of options it is felt necessary for immobilizing the car.

A small plastic box is needed to house the unit with either the PCB mounted socket fitted as shown or with a panel type fitted to the box end wall. The matching plug and length of six core cable is used to connect another panel socket mounted on the dashboard near to the ignition switch.

Another plug wired to supply Q1 and Q2 (the 'key') when inserted into this socket is kept on the owner's keyring. At least three 'keys' should be made with one kept as a spare. The wiring combination is done to the constructor's choice.

Use black single-core flexible cable with a size of 0.75 mm csa (cross section area) for the control wires and keep the pairs of wires twisted or taped together to avoid the circuits getting mixed up. The six core cable can be 0.2 mm csa stranded the same as the supply wires.

To mount the panel socket on the dashboard a 16 mm hole (a size which should be checked) is needed and should be close to the ignition as mentioned earlier. Push the wired-up plug and the cable through the hole until the socket sits neatly into position. Two 12 mm long No 6 self-tapping screws used as fixings will complete this part of the installation.

Plug in the lead from the dashboard socket to the socket fitted to the plastic box. Locate the box into a suitable position under the dashboard away from moving pedals, etc. Run the control cables to the various destinations required for their functions making sure the cables are tidy and not dangling around or likely to catch on foot pedals or other moving parts. Take care to keep wires away from the hot parts of the engine when routing the ignition control cables. It may be useful to consult a workshop manual to locate the cables which need to be connected into. Connections should be soldered and then insulated with PVC tape.

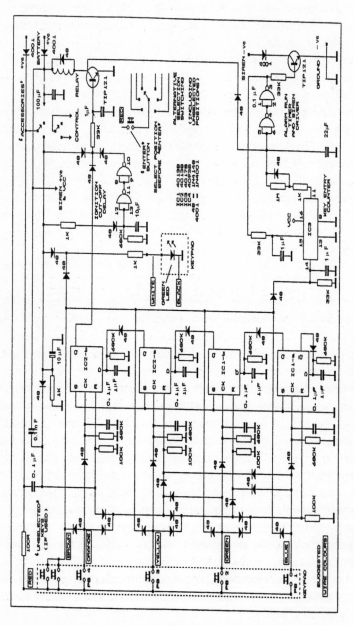

Figure 1.9 Four digit keypad immobilizer (circuit)

Next, drill a small hole in a hidden position for the chassis connection. Use a crimp-on eyelet and a small self-tapping screw to ensure a good earth. The supply must be one which is present when the ignition is on but does not disappear when the engine is cranked. Put an in-line lamp fuse in this wire near to the supply connection.

If a sounder is required, then fit this on the passenger side of the car under the dashboard, again keeping the wires tidy and out of sight. When the ignition is turned on the sounder should immediately sound and stop as the immobilizer 'key' is inserted. With the ignition on and no 'key' fitted, none of the controlled functions will work. Inserting the 'key' should then return everything back to normal.

It may also be a good idea to fit the unit one step at a time, testing the last connected function separately.

An effective but unusual system can be fitted using this immobilizer and the flashing beacon. This is certain to make it hard for the thief who knows commercial systems.

A coded keypad immobilizer system

To overcome the possibility of keys or other car-enabling devices being lost or stolen, and the car being driven away as a result, another method can be used to disable a vehicle other than those already described. The 'coded' immobilizer is a simple-to-use alternative which needs both the ignition key and the correct code input for starting the car. The full circuit for this is shown in Figure 1.9.

A minimum of four code pushbuttons are needed but other unselected keys may be added. These are fitted to a small panel which is mounted in a convenient operating position for the driver. The unselected keys only operate the keypress counter adding to the total keypress inputs. The selected code sequence buttons are used to increment both the code and keypress counters. A total of nine keypresses will trigger an alarm run timer which should run for about 30 seconds. The correct code will enable the car ignition and immediately stop sounding any alarm call due to an incorrectly entered code sequence.

An LED mounted on the keypad illuminates during a keypress and gives a steady glow when the correct code has been entered and the car is ready to be started.

Circuit description

The ignition must be turned on to enable the keypad and simultaneously reset both the sequence and the keypress counters. IC1 and IC2 are 4013 'd' type flip-flops and together they form the code sequence counter which is incremented as the correct keys are pressed in the right order.

When PB1 is pressed the data (logic 1) present on the D input is clocked through and is reflected at the 'Q' output.

A diode matrix connected around the pushbutton inputs will always reset the counter flip-flops when an incorrect key is pressed. This ensures that only the last four keypresses in the correct order will operate the relay driver.

Construction

The PCB drawing and component overlay are both shown in Figure 1.10. It includes transistor base diagrams and fitted diode orientations. Take care when fitting the 4017B Ics (integrated circuits) and double check the diode polarities as you fit them.

The wiring diagram in Figure 1.11 describes the wiring to the PCB and keypad assembly. A suggested method of constructing the keypad is shown complete with the necessary interconnections.

Follow as far as possible the wiring colour code suggested for the keyboard, this will help if a change of code is needed or if there is a fault to be traced. Use the wire types and sizes used on earlier immobilizers for the supplies and control wiring, taking note of the instructions on car wiring practices.

Keyboard components will vary according to suppliers' stocks, but a minimum of four buttons will be needed. Buttons cannot be used twice but the addition of extra unselected buttons will give added security. Arrange the button wiring in a random connection pattern rather than have them in any logical sequence and mix the used keys with the unselected ones. A standard matrix keyboard is not suitable for this application, but other types with separately connected buttons may be available and these would present a more professional look.

Another approach would be to use a single pole rotary switch with 6

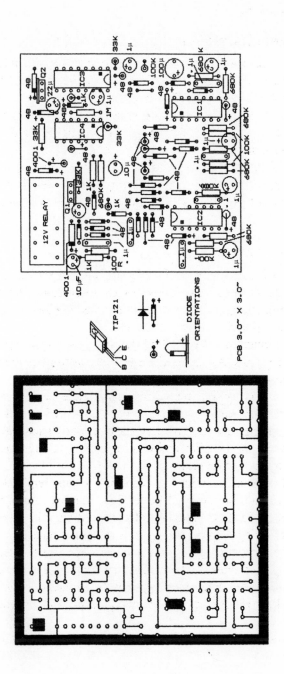

Figure 1.10 Four digit keypad immobilizer (PCB and layout)

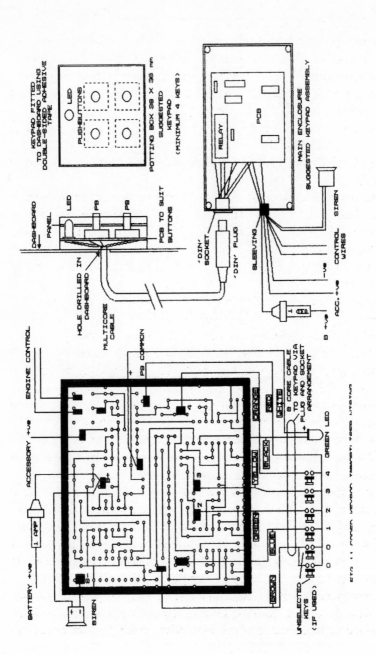

Figure 1.11 Four digit keypad immobilizer (wiring and assembly)

or 12 positions. This would then be rotated to the positions in sequence and pressing an 'enter' button after a number is selected. The 'enter' button would then supply a logic 1 to the selected flip-flop in turn until the sequence is complete. A numbered control knob, pushbutton and LED would then complete the panel.

The unit can be used to control other devices such as the alarm systems to be described later, or an extra relay can be added to extend the disabling capabilities by disconnecting the fuel pump and starter solenoid, etc.

This output is connected to the 'D' input of the second stage via a diode where a 1 μF capacitor is connected in parallel to ground to hold the logic state. When the second key is pressed (PB2) the first stage flip-flop is reset and the 'Q' output of the second stage goes high to reflect the data input present at the time. The data input then returns to a logic zero as the capacitor is discharged.

As the buttons are pressed in order the counter increments until the 'Q' output on flip-flop 4 is high at logic 1. The high output drives the base of a TIP121 darlington transistor relay driver switching it on and energizing the control relay. One set of contacts is used for engine control and the other set supplies the TIP121 base causing the relay to latch until the accessories supply is turned off. On most cars the accessories supply is lost during engine cranking, a timer built around two schmitt nand gates retains the base supply for about 10 seconds. This is long enough to enable the car to be started. The delay is very useful if the engine is stalled and needs to be restarted quickly without the need to re-enter the system code.

The keypush totalizer counts the number of inputs and will sound the alarm on the ninth push. A decade counter is used and this drives the input of a 30 second timer formed around the two remaining gates of IC 4 when the output of the counter reaches nine. The timer output switches on a second TIP121 darlington transistor which is used to provide a current sink or negative supply for driving a siren.

The sequence counter is held under 'reset' by taking the reset inputs high with the output from the latching contacts of the relay. A diode connected to the collector of the relay driver discharges the alarm call timing capacitor causing the output to switch off the siren when the relay is energized.

By using a fixed code immobilizer the problems of retaining a volatile memory is solved, if an immobilizer was used which required a back-up supply then consider the confusion which could arise if the car battery was flattened. This could happen if the vehicle was left unattended with the owner on holiday. If the memory retention battery was faulty and developed a limited life span and the code was lost, it is possible the driver would be unable to remember any default code. The driver may not even realize what was happening and could become totally confused.

These problems would not occur with a fixed code immobilizer because the memory is 'hard wired' and cannot revert back to any default code setting. The user can always rely on the code being correct and the unit ready for use.

Chapter **2**

The basic alarm system

Using the alarm

The basic alarm is easy to use and is controlled by using the ignition
key. When the ignition is turned off an exit delay is initiated which
inhibits all alarm trigger inputs. This is shown by a red LED glowing
steadily indicating the system status. After 45 seconds the red
LED starts flashing and becomes a warning beacon. The alarm is now
'armed' and ready to detect any intruders opening the boot, bonnet or
doors, etc.

When the driver returns to the car and opens a door he will trigger an
entry delay of a few seconds before the alarm is sounded. To stop the
alarm from sounding it will be necessary to turn on the ignition or the
car's accessory circuits with the ignition key.

If the alarm is not switched off within the entry delay period then
the alarm will change its operating mode and:

- use of the ignition key to control the alarm will be denied;
- the engine will be disabled;
- the alarm will be sounded for about 30 seconds.

Once the alarm has been sounded, the ignition control relay latches
and holds the alarm in its new operating mode.

To return the alarm to normal operation the relay must be reset. This
is best achieved by using a toggle switch to disconnect the alarm positive
supply and de-energize the relay. The switch should be hidden from

view–but still remain accessible to the driver–and its location kept secret. By using the switch as a service by-pass, the car can be serviced by the local garage without causing any fuss.

A green LED is provided for checking the alarm trigger inputs and when mounted close together the green and red LEDs form an alarm status display for driver information.

The green LED will glow whenever a door or boot, etc. is left open or if the accessory loop is broken. When the driver vacates the car, a quick look back at the LEDs will show if the car is secure. This function is especially useful for checking the trigger switches without the alarm being sounded or opening individual doors, etc. to prove their operation.

System control sequencing

The basic alarm system operates sequentially with an operating cycle which is dependent on its inputs for information on the car's security status. A flowchart has been included in Figure 2.1 to help understand the alarm operation.

The cycles for sounding the alarm are different for instant or delayed alarm calls should the car be left unsecured. A car is said to be unsecured if a door, etc. is left open. The cycle for an open door is as follows:

- 8 seconds entry delay;
- 30 seconds recycle delay;
- 5 seconds recycle delay;
- 8 seconds entry delay;
- 30 seconds alarm call.

Unless the alarm is turned off at the hidden switch the alarm will continue to sound in this cycle until the door is closed, stopping only after the alarm period has ended.

A similar cycle would occur if the boot or bonnet, etc. had been opened but would not include the entry delay. The alarm call would follow this cycle instead:

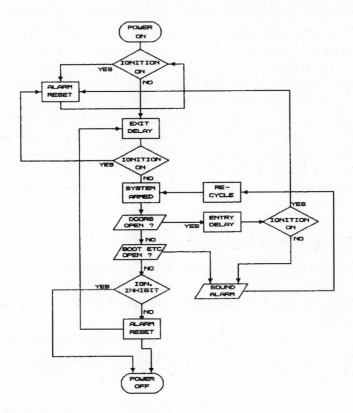

Figure 2.1 Basic alarm system (flowchart)

- 30 seconds alarm call;
- 5 seconds recycle delay;
- 30 seconds alarm call.

Stopping only after an alarm call if the boot, etc. was later closed. The operating sequence without an alarm call is shown below:

- 45 seconds exit delay;
- 8 seconds entry delay;
- ignition on.

Circuit descriptions

Four monostable multivibrators are used to provide the timing period required for the alarm. The 555 timer is again used and its operation in this mode is outlined below.

Figure 2.2 shows the basic circuit for a monostable using the 555 including the timing elements C1 and R1. Q1 is again used to control the i.c. by pulling pin 4 down below the reset value of 0.6 V. The full circuit for the basic alarm is described in Figure 2.3 and lists all of the monstables used in this design.

If Q1 is switched off, pin 4 is allowed to reach VCC enabling the timer to be ready for operation. With pin 4 at VCC and pins 3, 6 and 7

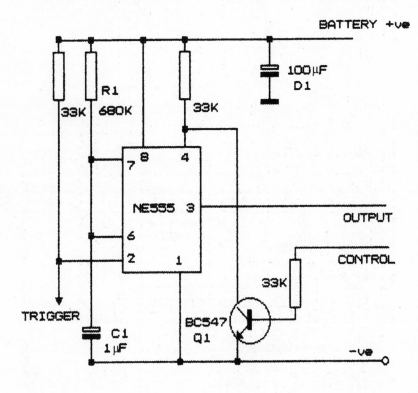

Figure 2.2 Monostable (circuit)

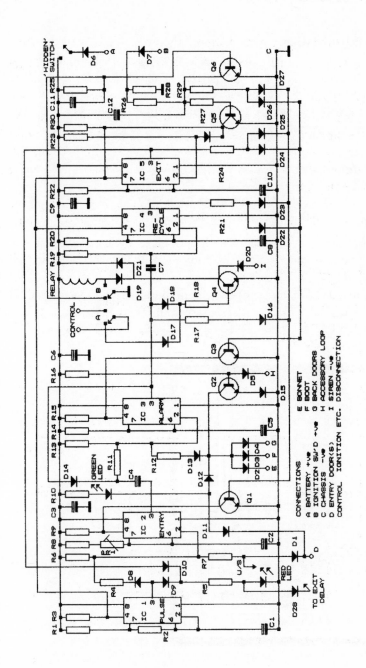

Figure 2.3 Basic alarm system (circuit)

CONNECTIONS

A BATTERY +ve   E BONNET
B IGNITION SW'D +ve   F BOOT
C CHASSIS -ve   G BACK DOORS
D ENTRY DOOR(S)   H ACCESSORY LOOP
                        I SIREN -ve
CONTROL IGNITION ETC. DISCONNECTION

low, a negative trigger pulse on pin 2 (below $\frac{1}{3}$ VCC) will initiate a timing period. Pin 3 will go high, pin 7 will release C1 and allow it to charge towards VCC via R1. When the voltage on C1 reaches $\frac{2}{3}$ VCC pin 6 will trigger and terminate the timing period. Pin 3 will go low and pin 7 will discharge C1. The timing period formula is:

t1 = 1.1 Rt Ct, where t = mSEC, Rt = Kohms and Ct = µF

example:

Rt = 680 k Ct = 47µF
680 × 47 = 31960 × 1.1 = 35.156 seconds

In practice, due to the actual values of electrolytic capacitors being greater than the stated values, the timing would in this case be around 45 seconds.

Figure 2.3 shows the complete circuit for the basic alarm and a PCB layout with component overlay is shown in Figure 2.4.

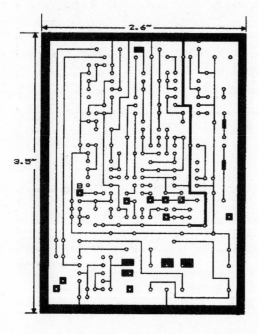

Figure 2.4 Basic alarm system (PCB)

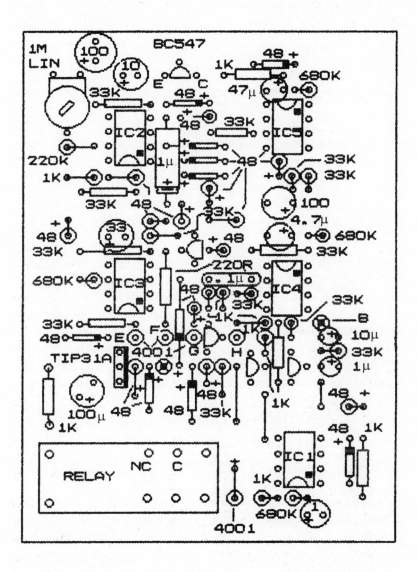

Figure 2.4 (cont) Basic alarm system (PCB and layout)

Listed below are the five main circuit sections and their control transistors:

Pulse generator   Q5
Exit delay   Q5
Recycle delay   Q5
Alarm call timer   Q3
Entry delay   Q1

The transistors are in turn controlled by their own various sources:

Q1   Ignition control: exit delay, recycle delay, alarm call
Q3   Ignition control: exit delay, recycle delay
Q5   Ignition control

The wiring for the basic alarm system is illustrated in Figure 2.5.

Exit delay

When the ignition is turned on, all of the alarm sections are held under reset conditions by their control transistors. This inhibits any unwanted operation of the alarm system while the car is being driven. An exit delay is required for the driver to leave the car giving ample time for the collection of belongings, etc. without sounding the alarm due to an open boot or doors.

There are two ways that an exit delay will be initiated:

* at power-up when the hidden switch is turned on;
* ignition off when switching off the ignition or accessory supplies with the ignition key.

When the alarm is first powered up C12 charges up via R27 and R29 (Figure 2.3). This action briefly simulates the ignition on and then ignition off conditions by resetting all of the alarm sections and starting the exit delay sequence.

When Q5 is switched off, either by power up or ignition off, pins 4 and 2 of the exit delay monostable are released and are allowed to rise toward VCC. As pin 4 rises to above 0.6 V pin 2 is still below the

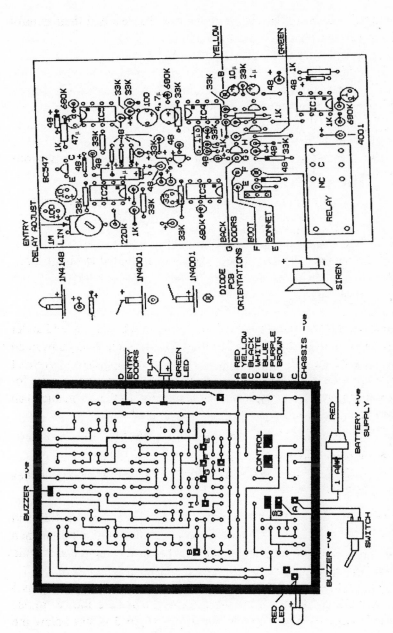

Figure 2.5 Basic alarm system (Wiring and assembly)

trigger voltage of $^1/_3$ VCC. In this condition the exit delay is triggered and the output from pin 3 now drives both Q1 and Q3 which holds both the alarm and entry delay timers under reset during the exit period.

On completion of the exit delay period Q1 and Q3 are released, all of the alarm sections are enabled and the alarm is now ready to detect any intrusion should any of the doors, boot or bonnet be opened.

The red LED should now be flashing indicating an armed alarm system, with the green LED off showing a secure vehicle.

The green LED

Power for the green LED is supplied by the resistor R10 which limits the drive current to around 12 mA. Diodes D11 and D12 isolate the delayed and instant alarm sections providing separate circuit paths for switching on the LED. The circuit is completed when a trigger diode is grounded through a door courtesy light switch or a specially fitted boot/bonnet alarm switch. The purpose of the trigger diodes is to isolate the alarm circuits from the car's electrical system and ensure only grounded connections affect the inputs.

Another trigger circuit is supplied and is used mainly for the protection of in-car entertainment systems such as car radios, etc. This accessory loop should be connected to the chassis of the equipment to be protected.

If the loop is broken, Q2 will be biased on by resistor R16 and its collector will trigger an instant alarm call by pulling pin 2 of IC3 below the trigger threshold. The green LED will stay illuminated and the alarm will sound in sequence until the wire is again grounded or the alarm is switched off.

The entry delay

By connecting the input diode D4 to the driver's door courtesy light switch, the entry delay will normally be operated by all of the doors connected to this circuit. It may sometimes be found that the driver's door is isolated.

When a door is opened, pin 2 of IC2 will be pulled down below the trigger threshold initiating the entry delay period. Output pin 3 will go high and discharge the coupling capacitor C4. C2 will start charging up via R8 and preset resistor PR1. If pin 2 of a 555 is kept below the $\frac{1}{3}$VCC threshold, then the IC will not terminate a monostable delay period. This problem can arise if a door on the entry delay circuit is left open. To overcome this situation an output from the pulse generator is fed into the bias network used for pin 2.

With the entry doors closed and the switches open-circuit pin 2 will be at VCC. With an open door D1 will be grounded but the voltage on pin 2 will be raised and lowered around $\frac{1}{3}$ VCC value by the pulse generator output. This allows the i.c. to terminate the delay period and retrigger the alarm due to the car being left unsecured. This would give an alarm call sequence for an open door as previously explained.

When the delay terminates pin 3 goes low and C4 starts to charge up again via R11 and R13. This produces a negative going pulse on pin 2 of IC3 which triggers the alarm call timer. The alarm call could have been prevented by turning on the ignition and resetting the alarm system before the entry delay had been completed.

The alarm call timer

IC3 is triggered and an alarm call is produced when a negative pulse arrives from IC2 or if one of the trigger diodes is grounded by a switch fitted to the boot or bonnet.

When an alarm call is produced, pin 3 of IC3 goes high. This output drives Q1 via R17 and D16 to hold IC2 under reset preventing it from being retriggered during an alarm call if a door is left open. The output from IC3 also drives Q4 via D18 and R18 which operates the relay and drives the siren. Pin 2 of IC3 is pulled up by pin 3 to ensure the alarm call will terminate.

Q4 is expected to operate loads of around 600 mA; this will include the main siren (at 350 mA for a 118 dB version or 450 mA for a high power type), the ignition control relay and the internal sounder.

When the relay is energized the set of contacts (A) change over latching the relay. The normally closed contact is used to ground the base of Q6. When the relay operates and the contacts change over, Q6 is switched on and its collector pulls down the base of Q5 to ground, preventing the ignition input from resetting the alarm. Relay contacts (B) disconnect the ignition supply and allow a hot-wired ignition to drive the siren via Q4.

These features are retained until the alarm is reset by switching off the supply either by using a hidden switch or a version of the 'electronic immobilizer'. At the end of the alarm call period pin 3 of IC3 goes low and allows C7 to charge up via R19. This generates a negative pulse used to trip the recycle delay timer IC4.

Recycle delay

When the alarm call has terminated IC4 is triggered to allow the alarm to be rearmed after a short delay.

With the introduction of the ultrasonic sensors (to be described in Chapter 4) it was found that a settling down period was necessary. This period of about five seconds temporarily holds the alarm and entry delay timers under reset. When the ultrasonic sensors are powered up an unwanted negative pulse is delivered to the alarm. Another consideration is the legal requirement for alarm sounding duration, a 30 second period is the maximum allowed and the recycle delay breaks up what may have been a continuous alarm call.

When the delay is triggered by the negative pulse from C7 the output pin 3 switches high driving transistors Q1 and Q3. IC1 and IC3 are held under reset but are released after about 5 seconds when the delay is complete. At this point the alarm will sound again if the car is left unsecure.

Accessory protection circuit

In addition to the normal trigger inputs, a normally closed circuit which can be used to protect any accessories fitted to the car. These can include fog lamps, radio and possibly even a trailer if necessary.

The base of Q2 is held low by the accessory loop and is biased high by resistor R16 if the loop is broken. D5 isolates the loop from external voltages and D15 compensates for the 0.6 V drop in D5 by raising the emitter to 0.6 V. This loop can be taken around the accessories in such a way that they cannot be taken without the loop being broken.

Conclusion

The instructions on construction and installation are followed, this alarm should meet BS6802 (part 2).

Although this alarm in its present state is basic, it will detect any intrusion into the car via the normal entry points. This makes it suitable for 'soft top' cars or other draughty vehicles which cannot use ultrasonic sensors.

The nuisance factor involved with this alarm is low. Vibration from passing traffic cannot affect it, only an opened door or other access point can cause an alarm condition giving the unit a high reliability rating.

The following chapters describe systems for those who require a greater degree of security.

Chapter **3**

The 'special' alarm system

To enhance the basic alarm system capabilities and provide extra protection for the car and driver, a few extras can be fitted easily to the basic alarm unit. Some of these extras will help the driver to use the alarm while others should make it harder for the car thief to take what he wants from you or your car. Figure 3.1 describes the PA (personal attack) button and internal sounder circuits, while Figure 3.2 is concerned with the courtesy light delay and the entry warning buzzer.

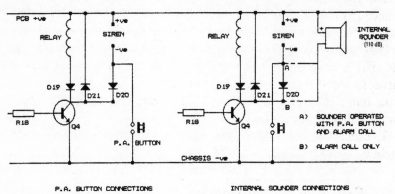

Figure 3.1 Personal attack button and internal sounder connections (circuit)

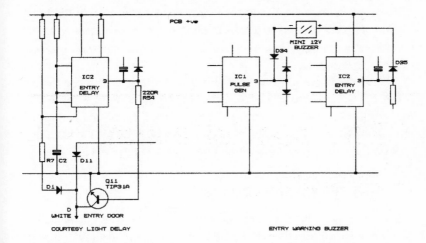

Figure 3.2 Courtesy light delay and entry warning buzzer (circuit)

Personal attack button

The ability to sound the alarm at any time and attract attention can be useful in certain circumstances. Women may feel more vulnerable at times and this feature may help to put them at ease.

The car engine must be able to continue running if the alarm is sounded by pressing the button and this is effected by wiring a latching push button between ground negative and the siren negative at the top of D20. By connecting the push button in this way, Q4 is bypassed and the relay is not energized allowing the car to be used. This allows the driver to sound the alarm and drive the car away out of danger at the same time. The use of a latching button retains the alarm condition until the danger has passed.

To cancel the alarm the driver only needs to press the button once again for normal operation.

Courtesy light delay-on

On dark and stormy nights this little circuit may help the driver jump into the car and find the ignition switch without a great deal of fumbling around. This reduces the risk of sounding the alarm and annoying the neighbours or undermining the validity of an alarm which is known to be forever sounding.

The entry delay output from IC2 is used to drive Q11 when the car is entered via the doors. This continues until the system is reset with the ignition key or the alarm is sounded. Q11 is connected across the entry door switches and when biased on will override the switches and turn on the courtesy light until the entry delay is terminated.

Entry warning buzzer

It is hard to estimate a time period and when the driver opens the car door he or she knows there is a time limit but may panic a little in case it takes too long to turn off the alarm.

A buzzer which gives a pulsed output can be used to count the number of seconds delay allowed for re-entry to the car. This facility will help the driver but give an intruder a sense of impending doom!

The buzzer must be of the miniature 12 V electronic type with a low current requirement, preferably less than 100 mA. Again the entry delay output is used for the positive supply and fed via D35. Use of D34 and D35 prevents any unwanted interaction between other circuits and ensures a correctly polarized supply to the buzzer.

When the entry delay is active the buzzer circuit is completed only when the pulse generator output is low. This grounds the buzzer negative every half cycle giving it an approximate 1 Hz output.

Internal sounder

When an intruder breaks into a car he or she may assume that an external siren will be sounded. To have an internal sounder creating a deafening wall of sound at the same time may be the last thing he or she expects. A sounder giving a 100 dB output in an enclosed space and no means of stopping it may deter any further intrusion.

The constructor has the choice of operating the internal sounder in two ways: either at any time the main siren is sounded, including using the panic switch; or only when the alarm is sounded by an intruder. By connecting the sounder in parallel with the main siren it will be operated on all alarm call occasions. If the sounder negative is connected directly to the collector of Q4 then it will only sound during an intruder initiated alarm call.

There are many sounders suitable for use in this situation and preferably should be of the high output piezo type with a warbling, high frequency, piercing tone.

Internal battery back-up

To maintain the alarm system control at all times it is necessary to maintain its power supply. The most suitable battery for this purpose is the lead-acid 'dry' battery.

Other types of battery could be used but have their own individual disadvantages. Dry cells will eventually discharge leaving the car owner with no back-up protection, plus there is the risk of dry cells leaking and causing damage to the car interior. 'NiCad' cells could be used but special arrangements would be needed for charging them and supplying power to the alarm.

With a 'dry' lead-acid battery, no special circuitry is needed other than to insert a diode into the main supply line from the car battery. This diode prevents the back-up discharging back into the car's electrical system. Charging is straightforward with the back-up simply being connected in parallel with the incoming supply.

In the event of the main battery being disconnected, power will be delivered to the alarm without any break in supply. All alarm conditions will be maintained, including the ignition control circuits. If the main battery loses charge for any reason while the alarm system is 'armed' then, if the voltage drops to below approximately 4 V, the alarm would be sounded. This is due to the trigger input being pulled below the $\frac{1}{3}$ VCC trigger voltage via the interior courtesy light. Disconnection of the main battery may not cause the same effect but would not stop the alarm from functioning properly.

Electronic key service/reset switch

The final modification to the original basic alarm is to add in an electronic switch in place of the mechanical hidden toggle switch. Perhaps the easiest way to accomplish this would be to use the 'electronic key' immobilizer as described earlier.

By using the common and normally closed connections of the relay contacts 'b' and disconnecting the sounder supply diodes, the alarm could be reset or put into 'service mode' to allow work to be carried out on the car without the alarm causing any interference to car mechanics, etc. Using the 'key' in this manner could also make it easier to load the car after a day's shopping! Needing an extra 'key' will give the owner added anti-theft security.

It would seem unlikely that an average car thief would try to overcome the combination of the key plug. Because of this the use of an extra relay operating in parallel with the immobilizer relay to give another two features as described in Chapter 1 may be used.

The combination of the alarm and the immobilizer can be carried to suit the individual constructor depending on his or her needs. Whether both 'keys' are needed to start the car or simply the electronic 'key' to perform the reset functions is a matter of personal preference.

Chapter 4

Ultrasonic intruder detection

Interior protection

To guard against an intruder gaining access to the car by breaking the window glass, it is necessary to have the ability to detect the presence of a moving object inside the car. There are two basic types of detector available, these are either 'passive' or 'active' in their operation with varying degrees of suitability.

Passive detection could be effected by using infra-red sensors similar to those used in household burglar alarms. Unfortunately these devices would be very susceptible to false alarm calls. Sunlight penetrating the car, clouds and external movement would all combine to cause false triggering of the alarm.

A simple light beam system with both sender and receiver waiting for a broken beam for intruder detection could be used. Setting up would be difficult because visible light would not be suitable, but infra-red would be invisible and could react to rapid variations in infra-red radiation from sunlight, etc. However, this could cause false triggering, and mounting mirrors and keeping them aligned over long periods would be another unwanted problem to the owner.

The use of microwaves is the most suitable method of detecting intruder movement. One major advantage is the reduced likelihood of false alarms due to air movement. Unfortunately microwave generation is not legal without either a licence or type approval from the Department of Trade and Industry. It is also relatively expensive and if poorly constructed could be dangerous to the user, especially if a fault allowed the system to become active while the car was being driven.

This leaves ultrasound as the best system for the home constructor to use in the car alarm system.

Movement detection using sound

Whenever a sound wave is generated in a continuous and regular fashion the space around the generator is filled with a field of steadily radiating sound waves. If the generator is placed inside an enclosed area the sound will be reflected back and forth slowly diminishing in amplitude. A fixed pattern of waves is set up inside the enclosure and in some places waves will meet in the same place creating stationary pressure peaks known as 'standing waves'.

Standing waves can have a pressure value of up to the sum of the two meeting wave fronts, this can lead to large differences in pressure between the peaks and troughs. If a moving object is placed inside the field of sound, the pattern of waves will vary according to the position and direction in which the object is moving. The frequency of the standing waves moving inside the area will be dependent on the speed and direction of the moving object in relation to the receiver transducer.

In a car, if two sensors are positioned above the front doors, pointing slightly downwards and inwards towards the gap between the two front seats, a signal from the transmitter would be reflected back up to the receiver by any moving object in the general area of the front of the car. Should the moving object be an intruder then the alarm would be triggered when the 'moving' sound ripples caused by the reflections are received during the alarm 'armed' condition.

The ultrasonic transmitter

For the generation of ultrasonic sound waves a transducer is used which employs a ceramic 'bender'. This material is a piezo-electric ceramic which distorts or bends whenever an electrical voltage is applied to it. If this is driven by an AC signal then it will vibrate and convert this into a sound output. The ultrasonic transducers are designed to be driven at their resonating frequency at which the transmitter output and the receiver input sensitivity will increase significantly but only on a narrow band around the centre frequency.

An accurate and reliable signal source is required to maintain the transmitter output at a constant level. This is achieved by using a quartz crystal controlled oscillator to give the stability needed for reliable operation. Frequency stability is excellent and under a very wide range of temperatures, operation is almost unaffected. Included in the diagram for the ultrasonic sensor system of Figure 4.1 is the circuit for an oscillator which uses the 4069UB cmos hex invertor IC. Invertors 1 and 2 are used to generate the 40 kHz signal with the quartz crystal in the feedback loop to control the frequency. The output of the oscillator is fed to the input of invertors 3 and 4. The bridge formation is completed by connecting the inputs of invertors 5 and 6 to the outputs of 3 and 4 with the load connected to the outputs of the two invertor pairs.

The drive to the transducer load is approximately 18 V rms which is slightly down from the maximum of 20 V rms recommended for this type of transducer.

The ultrasonic receiver

For this circuit an LM3900N quad operational amplifier is used. This amplifier is different to the more common types which give an output that is proportional to the difference between the voltages applied to the two inputs. This amplifier is of the current differencing type known as a Norton amplifier, where the output is proportional to the current flowing in the two inputs.

Containing four op-amps, the i.c. performs the four functions needed to process the incoming signals.

- input amplifier;
- signal filter;

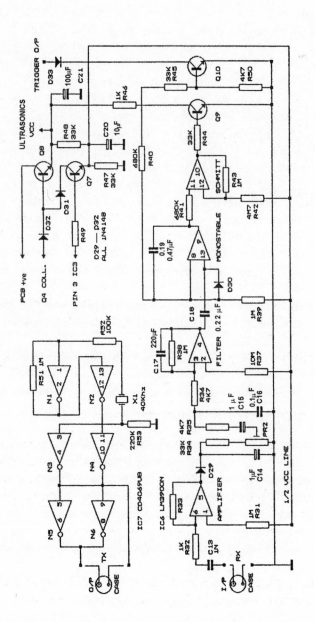

Figure 4.1 Ultrasonic motion detection (circuit)

- monostable;
- Schmitt trigger.

With an open loop gain of 70 dB for each amp and the ability to operate on a single ended supply from 4 to 36 V, this i.c. is especially suited for this application.

For linear operation on a single ended supply these amplifiers require one of the inputs to be biased at $\frac{1}{2}$ VCC. A resistor–capacitor network is used to give the decoupled half supply needed for correct operation of the receiver.

The input amplifier

With the feedback resistor and the half supply resistor at the same value (1 m) the output settles at a mid supply value and any input variation is reflected at the output with a gain determined by the input resistor R32.

The input signal is applied to pin 6 via coupling capacitor C13 and input resistor R32. The gain of this amplifier is set at about 1000 and is calculated with the following method, using the ratio of R33/R32 (1 m/1 K).

The output from pin 5 is rectified by D29, smoothed by C14 and appears across fixed resistor R34 and preset PR2. The setting of PR2 determines the overall gain of the receiver.

The signal filter

Input capacitor C15 couples the output to the low pass filter formed by R35 and C16. After the filter stage only the low frequency content of the signal from the first stage will reach the input of the third stage. C17 is included to provide stability and prevent oscillation which can be caused by capacitance in the input circuit.

The cut-off frequency (fc) at 3 dB down is determined by the formula:

$$fc \; 1/(2 \times 3.142 \; R36 \; C16).$$

The values used give a 3 dB cut-off at around 330 Hz and a rejection of 6 dB per octave or about 20 dB per decade, (the unwanted 40 kHz content of the receiver will be around 80 dB down on the wanted low frequency signals).

The monostable

After the signal processing has been completed by the first two stages the signal consists of many short duration pulses. These pulses need to be lengthened and to accomplish this a monostable is employed in the third stage.

When the non-inverting input on pin 13 receives pulses from the preceding stage, the output on pin 9 goes high and charges up C19 and pulls up pin 8. When the pulses cease C19 starts to discharge and releases pin 8 with a delay of about two seconds. D30 limits the differential voltages between pins 8 and 13 to 0.6 V to protect the inputs.

Schmitt trigger

Signal processing is completed by the use of the fourth stage, pulse-shaping schmitt trigger. The rise and fall times of the output pulses from the monostable are too slow for correct operation of the alarm.

By arranging the final amplifier in the inverting mode with a feedback resistor to the none-inverting input, a pulse on pin 11 will drive the output low, pin 12 will be pulled low resulting in an increased rate of change at the output. As pin 10 goes low Q9 is switched off allowing Q10 to turn on. When the collector of Q10 goes low D33 is forward biased pulling down pin 2 of the entry delay, triggering the alarm and switching on the green LED. Q8 is used to switch the power supply to the ultrasonic section. Diodes D31 and D32 pull down the base when the alarm is sounding and when the alarm is under 'reset'. By switching off the ultrasonic section during an alarm call, echoes from the siren cannot retrigger the alarm if they last longer than the recycle period.

Chapter 5

Construction

This chapter deals mainly with the construction of the special alarm but the methods used can be employed if the less complicated systems are attempted first.

Making the printed circuit board

Making the board by photographic means is not within everyone's pocket and in this case the board can easily be made by hand. This section deals with the drawing and etching of circuit boards using simple and basic methods.

When purchasing the copper clad board try to obtain the type with a base of fibreglass. This type is easily cut to size with a pair of shears or even strong scissors. SRBP is the other type normally available but this is a brittle material and may crack on the edges. To avoid this the board may be cut to size with a fine tooth saw or oversize with shears before trimming. When the board has been cut ensure that it fits into the box to be used. This saves trying to trim the board later after the tracks have been drawn.

The circuit board layout can be used directly as a drilling template, but a better idea would be to have it photocopied and to use this instead. Cut the copy as shown in Figure 5.1 and fit it to the board, trackside out, over on the copper clad side. Fold the 'wings' back over and tape them down tightly so that the paper drawing outline matches the board and is unable to move around.

The general hole size is 0.8 mm, but for some components, such as 1N4001 diodes, preset resistors, TIP31A transistors and the relay, the holes should be 1 mm. A drilling machine of some sort is required. A proper PCB drilling machine is desirable but a model maker's hand held type will do the job. If a hand held machine is used care must be taken to keep the drill vertical. Drill the holes directly through the paper template and try and be as precise as possible with the hole positioning.

When the board has been drilled, take off the template, and using a small, fine file, gently clean off any burrs from the holes. The board must now be cleaned with either fine wire wool or some other material such as a plastic kitchen cleaning pad. When the board is cleaned and dried, and your hands have been washed thoroughly, the tracks can be drawn on the copper surface ready for etching. It is essential that the copper surface is not touched with anything dirty or even slightly greasy because the etchant will only work on clean copper.

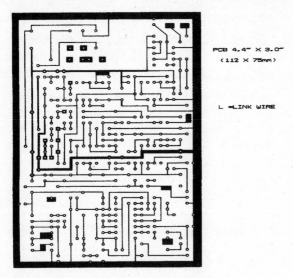

PCB 4.4" X 3.0"
(112 X 75mm)

L = LINK WIRE

Figure 5.1 Full special alarm system (PCB)

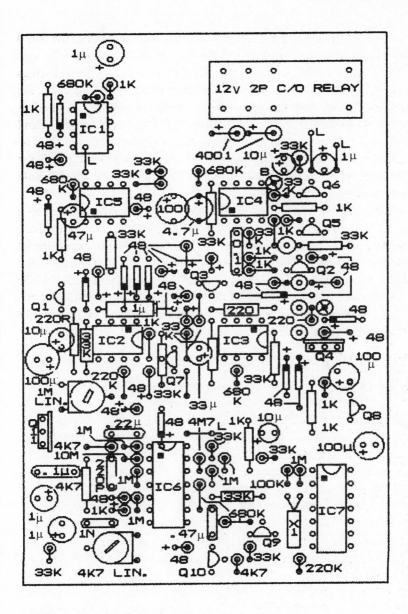

Figure 5.1 (cont) Full special alarm system (Layout)

Draw the tracks using the holes as guides and compare your lines against the master in the book. When the tracks are completed check for faults and correct these before etching.

After about fifteen minutes the etch resist ink will be dry. Turn the board over, and using a piece of PVC tape make a 'handle' on the back of the board so that it can be picked up without wetting your fingers with etchant.

Ready to use etchant (ferric chloride) can be purchased in a bottle and only needs to be poured into a plastic flat-bottomed container, such as an old ice cream carton. Place the board trackside down and float it on the surface of the etchant. Take it out again and check for any air bubbles on the board, bursting them if necessary, then replace as before. Check the process after about thirty minutes and if it is ready take the board out of the etchant and flush with cold water. After drying the board, clean off the ink with either fine wire wool or a plastic kitchen pad (do not scratch). Inspect the board for any unwanted fine copper lines, bridges or even breaks in the copper tracks. If all is well the board is ready for use.

Fitting the components

When fitting the resistors and diodes horizontally keep the body close to the board surface. When fitting them vertically, bend the top wire close to the top end then fit it so that the body at the bottom end is touching the board. The BC547 transistors need to be about 5 mm above the board and the TIP31A needs to be about 10 mm above the board to clear other components. The capacitors will vary according to their position and some components are fitted underneath them, but, above all, keep the leads short and neat to prevent accidental short circuits.

Start by fitting the wire links as shown on the overlay. Then fit the five NE555 i.c.s taking care to align them as shown. Next, fit all horizontal 33 resistors then the 1 k resistors and 1N4148 diodes. With all of the horizontal resistors and diodes fitted, install the vertical ones then continue with the BC547 and TIP31A transistors. The capacitors are next then, finally, the 1N4001 diodes, the preset resistors and relay.

All of the components should be fitted first with their leads bent over on the trackside of the PCB then cropped to about 3 mm or less in length. Good soldering is essential; dry joints and tracks soldered together can either cause, at best, very strange results or overload components which may need replacement.

When all of the components have been fitted, check for:

● component type;
● correct polarity;
● correct value.

With the exception of the supply polarity diodes, all of the 1N4001 diodes are fitted in a single hole with the free lead cropped to 3 mm length. This lead is used to connect the input from the ignition and the sensors to the wires used to connect the alarm to the car electrics.

The diode at connection B has its cathode connected to the PCB and the other diodes for connections D, E, F, G and H all have their anodes soldered to the board. If these are incorrect the function they relate to will not work.

At this point the components fitted should not include those for the ultrasonic section. If it is desirable, the completed alarm section can now be tested before fitting the remaining components and after the supply wires and LEDs are temporarily fitted reference should be made to the testing section near the end of this chapter.

When fitting the ultrasonic components start with diode D32 which is sited partly under IC6. Do not fit IC7 until last and do use an earthed soldering iron. IC7 is susceptible to static charges and can be destroyed if care is not taken to minimize this risk of static build-up.

Again check all components and connections and polarity, etc. before starting the next stage.

Preparing the enclosure

Three drills are needed and approximate sizes are 12 mm for the main cable entry, 6 mm for the ultrasonic transmitter and 3.5-4 mm for the receiver and personal attack button socket. By referring to Figure 5.3

drill the holes as shown then deburr them with a sharp knife. Two holes may be needed in the lid to fix the box into position during installation. These may be 6 mm and should in any case be large enough for the ties to be used.

Wiring up the circuit board

Start by connecting the sockets for the ultrasonics and the personal attack button to the board. Use, where possible, the correct colour coded wires as shown in Table 5.1.

Next, gather together all of the wires and cables listed in Table 5.1 while referring to Figure 5.2. These cables and wires should be selected by size and colour then cut to length as required or the lengths given may be suitable for most cars.

Table 5.1 Cable and wires required for 'special' alarm system

No.	Colour	Type	Size (csa) (mm)	Length (m)	Description
1	Red	Single	0.2	1.5	Main battery
2	Red	Single	0.2	1.5	Back-up battery
3	Black	Single	0.2	1.5	Chassis negative
4	Yellow	Single	0.2	1.5	Accessories/ignition
5	Green	Single	0.2	1.5	Radio cassette chassis
6	White	Single	0.2	1.5	Front door switch
7	Blue	Single	0.2	3.0	Bonnet pin switch
8	Purple	Single	0.2	5.0	Boot pin switch
9	Brown	Single	0.2	3.0	Rear door switch
10	Black	Twin	0.2	3.0	External siren
11	Black	Twin	0.2	2.0	Internal sounder
12	Black	Twin	0.2	1.5	Hidden/'key' switch
13	Black	Single	0.75	3.0	Ignition control
14	Any	Four core	0.2	1.5	LED display

Thread the wires through the 12 mm hole in the box and, after stripping them, solder the appropriate wires to the pads provided on the trackside of the board, double checking for the right wire in the

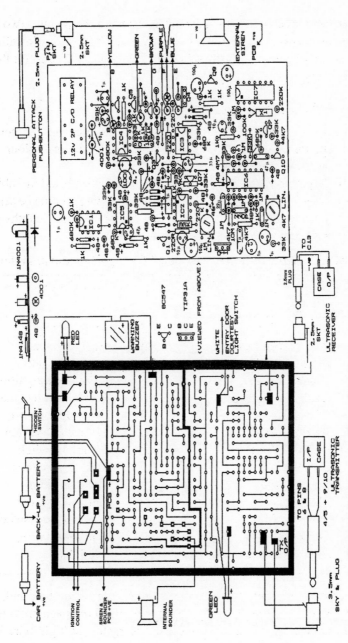

Figure 5.2 Full special alarm system (wiring)

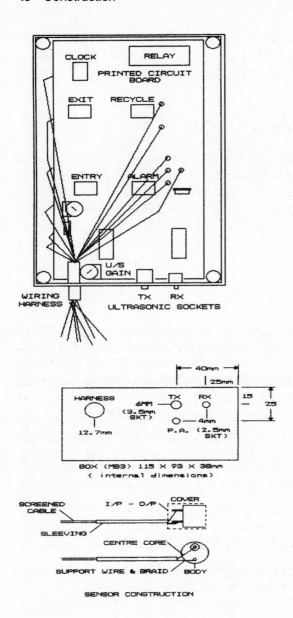

Figure 5.3 Full special alarm system (assembly)

right place as you go. When these are complete, turn over the board and complete the connections on the component side. Now connect and fit the sockets for the ultrasonic sensors and the PA button as shown in Figure 5.3

The PA button

A latching button is needed for PA use and would be best if it were coloured red to stand out on the dashboard. Use a 2.5 mm jack plug to match the socket fitted on the alarm box. A short length (1.5 m) of twin 0.2 mm cable will be needed for the lead. Take care not to short-circuit the plug connections with solder or by overheating the plastic insulation. Insulate the switch leads with PVC tape then check its operation with a test meter.

The LED display

Any stranded four core cable may be used but the best type would be that used in alarms of telephone applications but should not be too thin. Cut the LED leads to about 4 mm long and if possible slip 12 mm length of 4 mm shrink tubing over the cable. Strip back about 50 mm of the outer cover and using a cigarette lighter shrink the tubing over the end of the stripped cover. Slip about 5 mm of PVC tubing over the red and yellow (or equivalent) wires then strip the ends and solder them to the LED anodes (red wire to red LED). Solder the cathodes to the remaining wires (red LED to black lead, etc.) and insulate the connections. Test on a battery or power supply with a series 1 k resistor to limit the current.

Making the ultrasonic sensors

The sensors are usually purchased in pairs and are marked to show which is the transmitter and which is the receiver. Somewhere in the part numbers will be a T for transmit or R for receive. For the transmitter use 3.5 m of single core screened 3.5 mm diameter cable and about 1.5

m for the receiver. Fit a 3.5 mm jack plug to the transmit cable and a 2.5 mm plug to the receiver cable.

Take the transmit sensor and solder about 50 mm of No 21 swg copper wire to the lead connected directly to the sensor body. This acts as a support for the sensor when it is installed in the car. Take the free end of the tx (transmit) lead and slip about 60 mm of 4 mm shrink sleeving over it then strip back 10 mm of the outer cover. Group the screen strands together and strip back 3 mm of the centre core. Solder the screen to the back of the sensor body and the centre core to the free input pin. Pull up the sleeving along the cable, including the copper wire, then using a cigarette lighter heat the sleeving to make it shrink. This should now be gripping the cable and copper wires together. Insulate the sensor body with either 20 mm shrink sleeving or black PVC tape.

When finished complete the receive sensor in the same way.

Testing and fault diagnosis

Connect the green (accessories) wire with the black (ground) to the negative terminal of a 12 V power supply.

If a milliameter (0–100 mA) is not fitted to the supply use a multimeter on a similar range. Touch the red wire to the positive terminal noting the reading on the meter. An inrush current will be noted followed by a steady reading of around 10–15 mA. If the reading stays high or steadily increases, disconnect the 12 volt supply and check the board for solder bridges, reversed capacitors, etc.

When the alarm is powered up, the red LED should glow steadily and, with none of the trigger leads grounded, the green LED should be off. While the red LED is glowing steadily touch all of the trigger inputs in turn. This should make the green LED glow. If the ultrasonic sensors are plugged in and pointing in the same direction, a moving object in the vicinity will also cause the green LED to glow. The same effect will occur if the green wire is disconnected from the ground terminal.

If all is well check that the yellow wire resets the alarm and turns off the red LED when it is touched to the positive terminal.

Allow the alarm exit delay to time-out then touch the white delayed alarm (entry doors) trigger wire to ground. The entry delay will now start and if the buzzer is connected then it should sound intermittently. Using the yellow wire at this point will stop and reset the alarm. If the entry delay cycle is completed then the alarm cycle will start. An LED connected to the siren output will glow and the relay will energize. At this point only, disconnecting the supply will reset the relay, re-applying power will return the alarm to normal.

Table 5.2 Fault diagnosis

Fault Symptoms	Possible Diagnosis
High initial current	Reversed components (capacitors, i.c., etc.), solder bridges, faulty components
Overlong exit delay	This applies to all timing capacitors. The capacitor is 'unformed', desolder it and touch the leads to the power supply terminals for a few seconds (observing polarity), discharge it with a 1 k resistor then refit
Ignition 'reset' does not work	Incorrect diode polarity, faulty transistor, incorrect resistors, energized relay, missing reset link to IC1
Trigger inputs do not light the green LED	Incorrect input diode/LED polarities
Instant alarm call on power-up	PA button operating
Instant alarm after exit delay	Accessory loop open-circuit, instant alarm input grounded, fault on pin 2 of IC3
After an alarm call the ignition still resets the alarm	The relay has not energized or faulty contacts, Q6 faulty
No response from the ultrasonic section	Faulty 40 kHz generator, sensor or associated wiring

	No output from the receive circuits, faulty sensor or associated wiring/plugs, etc.
No output from generator	Damaged 4069UB due to static, very low output from the oscillator (can sometimes be remedied by reconnecting the sensor screen to ground and the centre lead to the original screen connection, check the plug and socket for continuity)
Faulty receive circuit	Check input plug and socket, set PR2 to half travel then test again by touching the input with a finger. If there is no response carry out the following tests with a 20 Kohm/volt multi-meter

1. Check for 12 V at pin 14.

2. Test for 12 V positive at pin 14 of the LM3900n.

3. Test output transistors Q9 and Q10 for operation.

4. With generator running test for AC output at pin 5.

5. Check for varying DC after D29 and R34 (touch input).

6. Pin 4 should switch high and low with an input.

7. C19 must not be 'leaky' (it should be open circuit to DC).

8. Test for $\frac{1}{2}$ DC volts at the junction R47–R48 (6 V).

With the ultrasonic sensors fitted and pointed in the same general direction as before, a moving object will now trigger the entry delay if the alarm is armed. Plug in and check the personal attack button; one push to sound the siren and another to stop it. An LED wired from the positive supply to the white wire will glow during the entry delay, indicating the correct operation of the courtesy light delay-on feature, but don't forget the series resistor to limit the LED current. If the receiver still does not work and all appears in order, then change the LM3900n and try again.

An oscilloscope would be very helpful at this stage if there are any faults. The quartz oscillator could be checked for voltage output and frequency and the receiver for stage by stage operation. If a frequency counter is at hand this could easily verify the generator output if necessary while still using the multimeter for the receiver.

I have experienced i.c.s breaking down after some hours of use, yet the vast majority work properly for several years.

When purchasing components try to buy as cheaply as possible but buy branded i.c.s to reduce the likelihood of premature failure. In the case of resistors and capacitors, buy only the power rating (mW) or voltage ratings, etc. stipulated in the parts list, this will save time and unnecessary expense.

If a fault persists there is only one answer, check and check again for polarity, value and type of components on the circuit board, but do not forget the wiring.

Chapter **6**

Installation

Fitting a complex car alarm can be a time consuming process. It can be expected to take up to $4\frac{1}{2}$ hours for a full 'special' system complete with all of the extras. On the day of fitting be sure you have everything you need because it may not be possible to buy or borrow them locally and you certainly can't use the car with wires, etc. on the floor!

Positioning the alarm box

The best position for the alarm box is behind the dashboard above the pedals. Using the car wiring loom for support is acceptable and it is possible to find an empty space where the box will fit. If such a place cannot be found then the box can be tied up using cable ties, etc. but it must be firmly secured; if it were able to fall it would cause the driver to lose control of the car.

Once a suitable place has been found, temporarily fit the box into position and then start routing the wires to their various locations. Start by finding access through the bulkhead between the engine compartment and the passenger area. It may be necessary to drill a hole. The utmost care must be taken to avoid any damage to the car wiring, If a hole is drilled, deburr it and fit a grommet to stop the wires chafing on the edges. Don't forget to keep the cables clear of hot or moving parts under the bonnet and clear of pedals and steering mechanisms, etc. under the dashboard.

Connecting the car's electrical system

In most cases the required connections can be found behind the car radio system for battery supply, accessories 'on' supply and the ground/ chassis negative. If the accessory protection loop is used it can be connected to the radio case.

After routing all of the wires to their respective locations, fit the LEDs in a prominent position then connect the accessory loop and chassis wires as needed. Using a voltmeter set to read 12 V, locate a supply for the yellow wire which is 'live' when the ignition key is turned to 'acc' or 'ign'. A permanent battery supply is also needed for the red wire.

Connecting into the car wiring should, for safety's sake, be done with the car battery disconnected when the main red wire is terminated and the ignition key is in the off position. Connect the ground/acc wires first then the ignition supply and the back-up battery, leaving the car supply until last.

Inline fuses should be fitted close to their power sources. Strip the insulation with a knife and solder the connections then insulate them individually with PVC tape.

Switch on! The red LED should now be glowing steadily. Check that turning on the ignition will turn off the alarm putting it under 'reset' and that turning off the ignition will return the alarm to an active condition. Leave the alarm 'reset' and start fitting the trigger switches.

Fitting the trigger switches

Beginning with the bonnet switch, find a suitable flat surface for the base with a connecting surface on the lid. Fit the switch using the instructions supplied and check its operation by making sure the green LED switches off when the lid is down and is on when the lid is up. Next, take off the driver's door courtesy light switch and feed the white wire through the body work. If there are two wires take off one and slip the bared end of the white wire into the connector body and re-connect. If it is the correct wire then opening and closing the door will operate the green LED. The last switch is the boot trigger. If there is a light fitted in the luggage compartment then take the purple wire to this and connect it to the grounded side of the lamp. This can be found by touching in turn the lamp connections, watching for the green LED or fit a switch to be operated by the boot lid.

Installing the ultrasonic sensors

Starting with the transmit sensor, pull down the edging strip from around the front of the door frame and fit the cable into the strip. Leave a short amount for the sensor and refit the strip into position with the sensor facing the space between the front seats. The best position for the sensors is just behind the sun visor in its reset position. Fit the receive sensor above the passenger door directly opposite the transmit sensor. Thread the cable back into the alarm box and with the power off plug in both leads. Switch on again and check the sensitivity of the ultrasonics, adjusting PR2 to suit the vehicle size. If the sensors are left too sensitive then high winds blowing through the dashboard vents may trigger the alarm. This must be considered when setting up the system.

Fitting the remaining extras

Only two extras remain to be fitted now: the internal sounder and the personal attack button.

Fit the sounder under the dashboard out of sight. Tie it to other wires or screw it to a solid surface keeping the wires tied up and tidy.

The connections may be soldered or mechanical, but must be well insulated.

The PA button is best fitted anywhere within easy reach for access by either the driver or front passenger. An unused switch blank may be used to save drilling the main dashboard.

Fit the switch according to its construction and plug it into the alarm box. Pressing it should sound the alarm when the system is powered up.

Testing the installation

Power up the alarm. While the exit delay is in operation check that all opened doors turn on the green LED. Check the boot and bonnet and then with the doors, etc. closed check again the ultrasonic sensors. After the exit delay has timed-out open a door and the buzzer should sound. Turn the key in the ignition to stop the alarm from sounding.

Turn off the ignition again and allow the exit delay to time-out, open the boot to check for an instant alarm call then try to stop the alarm using the ignition key. The siren should not stop and the engine will be impossible to start.

Reset the alarm by switching it off then on again; everything should return to normal. Any faults at this stage must be rectified and reference to the fault diagnosis section may help.

With the alarm on but under 'reset', start the engine and switch on in turn all of the car accessories to make certain none of them interfere with the alarm operation or vice versa.

With a little ingenuity, different sections of the alarm may be put to other uses. Care must be taken not to endanger either your life or others, by thoroughly testing any modifications before use.

Chapter 7

A multi-purpose garage or outhouse alarm system

When a car is left unattended in a lock-up or garage it still needs protection. Presumably the vehicle security system is switched on, operational and ready to sound the alarm? Unfortunately the car would also be out of sight, hidden behind a wall of brick or wood thus allowing the determined thief perfect cover to spend as much time as he needs to gain access to the vehicle. Sometimes the car may be out of action, or undergoing repairs, when the battery could be disconnected and therefore the alarm would be non-operational. It would then be possible for thieves to enter a lock-up or garage and strip the car of vital components at will.

To properly safeguard the car when garaged and to protect other items which may be of interest to an intruder, a self-contained garage alarm could be installed. This would be operated in a similar manner to a basic security system such as used on houses, etc. This unit could be used on its own or as a back-up system to the household alarm. An added bonus is that the stand-alone alarm to be described may be left on when it is impractical to use the main system. This could be during the evening when the occupants may be watching television, etc. Using this alarm for protecting the garden shed could save the user many pounds in gardening equipment or tools, common targets for thieves or burglars.

By utilizing its own internal dry batteries the unit to be described can be operated anywhere including garages, garden sheds and outhouses, etc. An alternative simple power source has been included for charging a 12 V nickel cadmium (NiCad) battery of ten AA cells for use when a mains supply is available.

As an added option, this alarm can be fitted to motor bikes, scooters, etc. and can be automatically controlled using the ignition system keyswitch. There is also provision for either normally open or normally closed inputs in the switch sensor circuits which should give a degree of scope when fitting switch sensors.

Circuit description

Two integrated circuits are used for this unit. These are CMOS types 40106B or 14106B hex inverting schmitt trigger and the 4013B dual D type flip-flop. The use of CMOS in this circuit is critical in terms of battery life. The quiescent current consumption of these circuits is very low and in the order of only a couple of microamps. The total current of the alarm system when switched on and 'armed' should only be a nominal 50 microamps. In some circumstances the battery life (using alkaline cells) can exceed twelve months of continuous operation.

The inverting schmitt trigger

This device is primarily used for wave shaping or as a means of 'squaring' a signal input. To do this two trigger levels are built into the input circuit. Figure 7.1 shows how these levels affect the output. One level known as the 'Upper Limit', which detects a rising signal, will cause an almost instantaneous high to low change in the voltage level at the output pin when the signal equals or exceeds that value. The other level known as the 'Lower Limit' will effect an upward change in the output when the signal applied drops to that value. This means that a varying voltage signal at the input terminal will create an output which simply switches between VCC and VDD (the supply rail

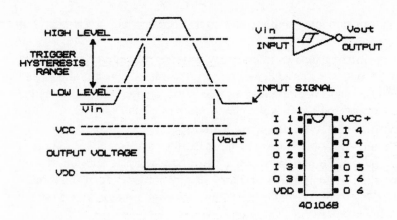

Figure 7.1 The schmitt trigger

voltages). The difference between the two trigger levels is known as the hysteresis value. This value and the switching limits will vary slightly between similar devices or manufacturers.

There are two different 40106B circuit configurations used in this alarm, the first is as a monostable used for the exit delay and alarm call timers. The second is an astable oscillator arrangement as used in the inputs.

The dual type D flip-flop

Abbreviations

Q	true output
Not Q	mirror image output
R	reset input
S	set input
CK	clock signal
D	data to be transferred

The D type flip-flop device is very similar to the set/reset flip-flop the outputs of which are controlled by the set and reset inputs. A high

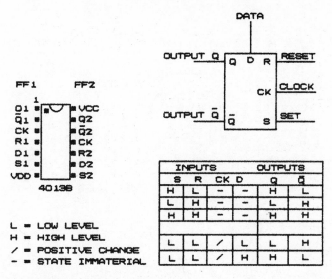

INPUTS				OUTPUTS	
S	R	CK	D	Q	Q̄
H	L	–	–	H	L
L	H	–	–	L	H
H	H	–	–	H	H
L	L	/	L	L	H
L	L	/	H	H	L

L = LOW LEVEL
H = HIGH LEVEL
/ = POSITIVE CHANGE
– = STATE IMMATERIAL

Figure 7.2 The dual D type flip-flop

logic signal applied to the set terminal (when the reset is low) will drive the Q output high and the Not Q low. If a logic high is applied to the reset input (when the set is low) the Not Q output will flip high and the Q terminal will go low.

The truth table for this is shown in Figure 7.2 and shows the addition of the D and clock inputs to the circuit. This allows the Q output to reflect the logic applied to the D input whenever the CK input has a transition from a low to high logic level. The truth table shows however that this is only true when the two inputs (R and S) are low.

Circuit operation

In Figure 7.3 the full circuit has been drawn complete with options for the battery/mains supplies and ignition input. A simple charging circuit for the NiCad battery is included but this is not intended as an alternative for a full mains driven supply. For full security the system should utilize a battery of some sort and not rely solely on mains power. If a power cut were experienced then the alarm would be dead and therefore useless.

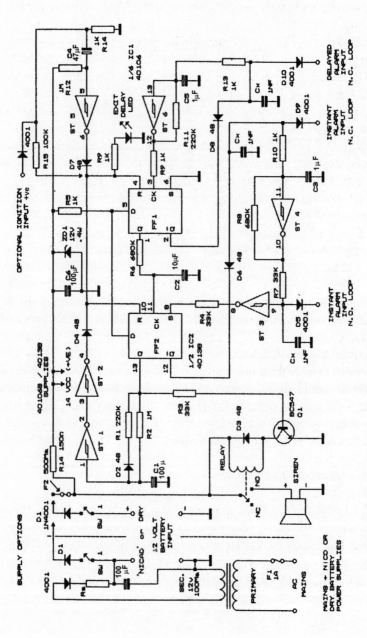

Figure 7.3 The garage or outhouse alarm (circuit)

Either use the dry battery option or the NiCad method as shown for full reliability.

When power is first applied the exit delay timing capacitor C4 starts to charge via R12 and R14. This pulls down the input of ST.5 which, due to its inverting action, applies a high output to the two reset terminals on IC2 (pins 4 and 10). A red LED connected to the reset terminals starts to glow indicating the exit delay is in operation. After approximately 45 seconds C4 reaches the upper switching level of the schmitt trigger input. At this point the output of ST.5 switches low releasing the reset condition. The alarm circuit is now 'armed' with both sections of IC2 now ready to accept inputs.

ST.6 is configured as a multivibrator, the values of R11 and C5 give a frequency of around 3 Hz. The delayed input at D10 is held low to ground via a normally closed loop switching circuit labelled as Zone 1. When the loop is open circuited or the switch contacts released, the voltage at the input of ST.6 rises as C5 is charged by R11. The charge on C5 increases until it reaches the upper switching limit of the trigger. At this level the output switches low. C5 now starts to discharge back through R11 until it drops to the lower switching level when the process will start again. This cycle is required to create a retriggering action, necessary if a door is left open thus simulating the opening of the circuit loop. If this facility were not available then the door could be left open during the exit delay and access to the area would not be prohibited.

The input built around ST.4 works in the same manner but has the additional inverter ST.3 in its path. When the circuit input at D9 is opened the output of ST.4 raises the input of ST.3 via R7. The output of ST.3 switches high operating the 'set' input of FF 2. Grounding the open circuit input at D5 triggers ST.3 in the same way but it must be noted that there is no retriggering action on this input.

FF 1 forms the delayed input latch and timer. As the output from ST.6 switches high it triggers the CK input. Due to the D inputs being at a high logic level the Q output switches high. Capacitor C2 starts to charge up via R6. This takes about 10 seconds to reach the CK trigger level. The data on the D input of FF2 is now clocked through to the Q output. The high level on pin 13 now drives a BC547 (Q1) transistor which energizes the siren relay. The other output known as the Not Q is normally a mirror image of the Q output. When the Q output goes high the Not Q is driven low. The input on D10 is now bypassed by D8 and will not now retrigger the CK input until the flip-flop has been reset.

The pair of inverters ST.1 and ST.2 form the alarm run timer which should sound the siren for a nominal 100 seconds. C1 is charged up via R2 until the upper trigger level of ST.1 at pin 1 is reached. Pin 2 switches low driving down the input pin 3 of ST.2. When the output of ST.2 is high it operates the reset inputs of both FF1 and FF2 then returns the Q outputs to a low level thus switching off the siren.

A recycle delay of approximately 7 seconds is effected when C1 is discharged back through the now Q output via R1 and D2. The input at D9 is bypassed during the alarm call and recycle delays via D6 to allow for retriggering of the instant alarm input.

The inclusion of three capacitors marked as C x 1NF (1000 μF) may be necessary when long cable lengths are used on the inputs to reduce the likelihood of false alarms due to static discharges, etc. This may be soldered directly to the PCB tracks with the leads cropped down to a minimum. It may be advisable to fit some insulation on the leads to prevent accidental short circuits across adjacent tracks.

Construction

CMOS devices are prone to failure when subjected to static charges. Most modern types have internal protection diodes on all inputs but care should still be taken when handling and fitting them. To help with this problem several methods have been used over the years. One easy to adopt measure is to securely earth the tools and workbench to be used. To do this you need to find a good ground connection. The best method is to drive a stake into the earth and take a wire from this to a sheet of metal fitted to the bench top. Alternatively take an earth connection from a cold water pipe. Don't forget to earth yourself–this can be managed by loosely tying a piece of uninsulated wire around a wrist and connecting it to your earthed workbench. As another safeguard your soldering iron bit should also be earthed.

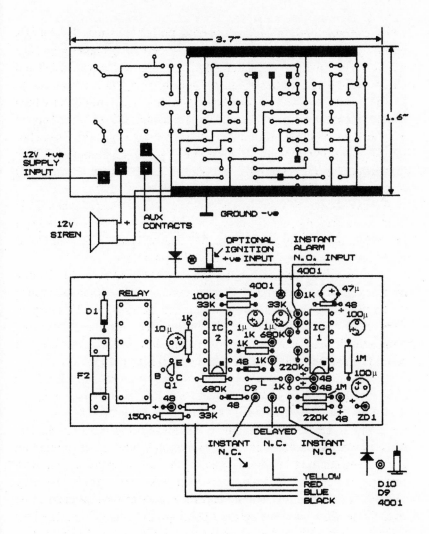

Figure 7.4 Garage alarm (PCB and layout)

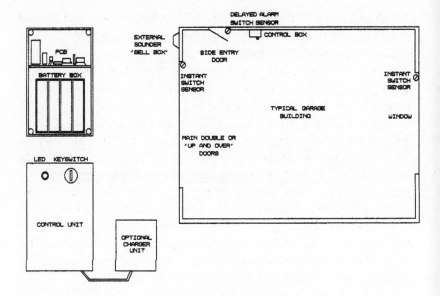

Figure 7.5 Control box assembly and typical installation

Fit the resistors and capacitors to the PCB first and then add the two i.c.s in turn, trying to keep your fingers away from the pins. Keep the soldering time down to a minimum to reduce heat damage. Fit the transistor and relay, then finally add the connecting wires.

The PCB should fit neatly into one of the slots provided in the box near the top. The battery box will then occupy most of the remaining space. Drill the lid to a suitable size for the keyswitch making certain that it will not foul the PCB or its mounted components. A smaller hole will be needed to accommodate the red LED, put this to one side of the keyswitch as shown in Figure 7.5.

Testing the PCB

When the PCB has been completed fit a series ammeter with a scale of 0–100 μA in one of the supply leads. Connect the ground or negative lead first then briefly apply the positive lead to D1 noting the current drawn. If after a couple of seconds the load is significantly high (above

2 mA) then check for faults. Check the PCB in the same way as described in previous chapters.

Ground the two NC (normally closed) inputs and again apply power. If the current is briefly high (this could be due to C6 charging up and the regulating action of Z1) but returns back to an acceptable level of around 2 mA then proceed testing the circuit operation.

The red LED will glow for about 45 seconds indicating the exit delay timer is operating. The LED will switch off when the alarm is 'armed' to conserve battery power. Open the 'instant alarm' input circuit and the relay will energize within one second. Turn off the power and back on again to reset the system then try opening the delayed input circuit. After a ten second delay the siren relay should again energize.

If the mains/NiCad option is used, be sure to test the voltage across the batteries and the charging current before connecting the PCB. This should be a nominal 25 mA continuous. Do not attempt to charge dry batteries especially alkaline types.

For dry battery operation do not install ZD1 or R15 as these will seriously affect the long-term battery life.

Installation

Mount the control box in an easily accessible position where it can be reached within the ten second entry delay.

If a separate power supply is to be used for charging the NiCad batteries then mount this externally but wire the 12 V DC output directly into the box as shown in Figure 7.5.

A typical sensor switch (a reed relay and magnet arrangement) is shown in Figure 7.6. The two components should be mounted on reasonably solid surfaces. Common positions include the door itself for the magnet and the door frame for the stationary switch. If a switch is to be fitted to double doors then the switch would be fitted to the fixed frame as before but the magnet must be fitted to the overlapping half which will open first. The two examples of wiring shown describe the methods used for installing the circuits. One is for the entry/exit door and is connected to the delayed input. This circuit uses the yellow

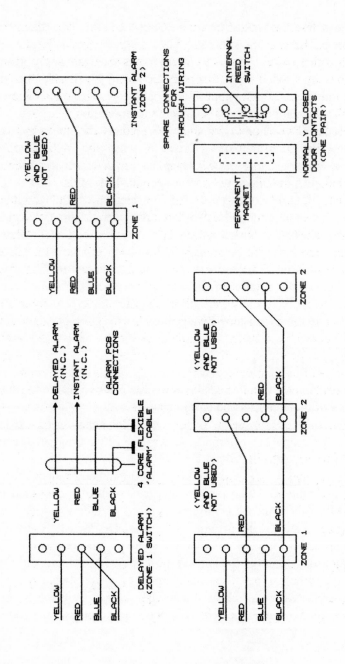

Figure 7.6 Switch sensor contacts (wiring)

and blue wires of the four core cable. The second circuit is the instant alarm loop. This uses the remaining black and red wires and is used on all of the other switch sensors. If a bell box is mounted on the garage then a beacon strobe may be fitted and connected in parallel with the siren to operate at the same time. This would give a visual indication of where the alarm is sounding. The cover bolt for retaining the cover will probably have provision for mounting a microswitch. This switch can be included in the instant alarm circuit to protect the box from being tampered with by an intruder. The switches can be wired separately of course but keep to the colour code to avoid confusion later.

After the alarm has been installed it is time to test the completed system. Close all of the protected doors and windows and switch on the alarm. The red LED should now glow to indicate the exit delay is operating. There should be no sound from the siren but if the instant alarm circuit is broken there will be a pulsed output to the sounder. This is due to the S input of FF2 being driven by the square wave signal from ST.4. If all is well, allow the exit delay to complete its cycle. If the entry door is now opened then the siren should sound after the 10 seconds entry delay is complete. A similar test on the instant alarm circuit (by testing each switch sensor in turn) should again sound the siren but this time without any delay. If a beacon strobe light has been included check that this is also working as expected. Check all connections and switches to ensure that they will not cause a false alarm due to wind vibrating the doors or windows they are fitted to.

Components list

resistors 5% carbon 1/4 watt

R1	220K
R2	1M
R3	33K
R4	33K
R5	1K
R6	680K
R7	33K
R8	680K

all radial electrolytic.

C1	100 µF	16 V
C2	10 µF	16 V
C3	1 µF	63 V
C4	47 µF	16 V
C5	1 µF	63 V
C6	100 µF	16 V
IC1	CD40106B	
IC2	CD4013B	

R9	1K	Q1	BC547
R10	1K	D1	1N4001
R11	220K	D2	1N4148
R12	1M	D3	1N4001
R13	1K	D4	1N4148
R14	150R	D5	1N4001
Box MB3 115 × 94 × 38 mm		D6	1N4148
Keyswitch single pole		D7	1N4148
4 × 2 (8) cell battery holder		D8	1N4148
5 mm RED+LED		D9	1N4001
LED+mounting clip		D10	1N4001
12 V DP. C/O relay		ZD1	12 V 400 mW
Single side PC board 3.7 × 1.6"		4 core multistranded	
		alarm cable	
3.5 mm cable clips		F1 PCB mounting fuseholder	
12 V electronic siren		500 mA glass fuse 20 mm	
		to match	

Options components list

10 size AA nickel cadmium cells
5 × 2 battery holder to suit
Transformer 12 V output @ 100 mA
1 x 1N4001
100 µF 25 V radial capacitor
Rs 200R 1/4 Watt carbon
F2 1 A glass fuse and fuseholder

Ignition control extras

100K resistor 1/4 Watt carbon
2 × 1N4001

Chapter **8**

Experimental circuits for remote control

For those of you who would wish to try remote control I have included some (hopefully) interesting circuits on this subject.

Radio or infra-red control?

There are problems with designing radio remote systems. The main problem is that not everyone has access to a frequency counter which can operate at 418 MHz for verifying the transmitter output. It is also unlikely that they will have a signal generator for aligning a radio receiver. The frequency that has been set apart for this use in the UK and Europe is 418 MHz.

Unfortunately there are different regions around the world where the radio frequency spectrum has been allocated or shared out differently amongst its users. To use a transmitter in a region for which it is not intended could result in causing interference to legitimate users.

Even building a simple transmitter/receiver type system could be hard for an 'RF novice' to accomplish. The receiver would have to be aligned properly and if it were a 'Super Regenerative' type then it may be unstable in use, unless the constructor has some experience of these.

The security of a radio remote controlled system is not as high as an infra-red version. The radio signal can be received, decoded and copied then the information can be used to program a special unit to disarm the vehicle to which it is fitted.

When constructing an infra-red transmitter the greatest problem that will probably be encountered will be the transmitter enclosure. These boxes (when sourced) could be vastly different from manufacturer to manufacturer. Electrically the units are fairly simple in terms of component counts. They do not generate any interference, the signal cannot be copied, and if the transmitter and keyring is lost then a potential thief will find it harder to locate the car by pressing the button and watching or listening for a response. To get a response from an infra-red remote system the transmitter must be actually pointing in the general direction of the receive diode.

Encoded infra-red transmitter

For security, an infra-red beam is modulated with a coded signal which is received and compared with a code selected in the receiver decoder. A simple to use integrated circuit is the UM3570 which can be used both as an encoder and a decoder thus combining both functions in the same package. The 100 K resistor and the 180 Pf capacitor form part of an oscillator. This, in conjunction with the grounded coding pins (1 to 12), generates the signal to be transmitted. The output from the encoder is fed to a PNP transistor which drives the infra-red emitting diode. The BC214L switches on when the encoder output switches low. The current limiting resistor will allow the infra-red LED to be driven with an approximate 50 mA load. Infra-red diodes are like other devices and have different operating characteristics. The device selected must not be overloaded at this value (Figure 8.1).

A short range receiver

The TDA 8160 is a purpose-made infra-red amplifier suitable for short range, low interference situations. The infra-red diode is internally biased automatically. The supply is only 5 V and needs regulating down to

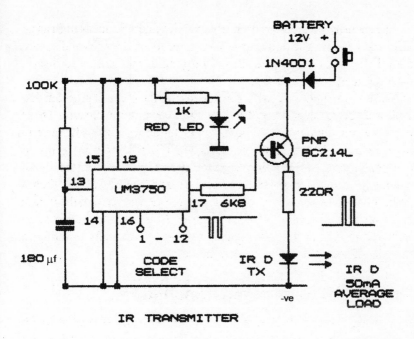

IR TRANSMITTER

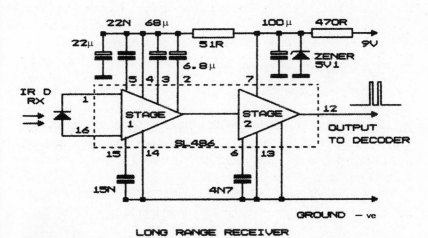

LONG RANGE RECEIVER

Figure 8.1 Infra-red transmitter and receiver (circuits) *Source:* Maplin Catalogue 1992 (Courtesy of GEC Plessey Semiconductors)

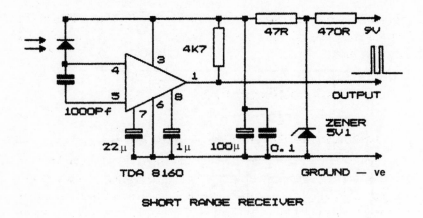

Figure 8.1 (cont) Infra-red transmitter and receiver (circuits)

this from the main supply as shown in Figure 8.1 with smoothing and interference decoupling capacitors. This circuit may be affected by direct sunlight swamping the infra-red diode making it most suitable for internal or dull situations.

A long range receiver

This circuit, built around the i.c. type SL 486, has good interference rejection. An AGC (automatic gain control) circuit allows it to operate in strong lighting conditions. With a three-stage amplifier and low frequency rejection filter it cuts out interference from street lighting. The infra-red diode is again automatically biased at the correct value for best results.

This receiver should be suitable for use in electrically noisy situations and at a greater distance than the low range unit. The infra-red diode must have some metallic screening around it soldered to a negative ground PCB track to reduce direct radio frequency pickup. This could be formed tin plate or even a piece of copper clad board cut and soldered into a U shape.

Output control circuit options

Remote on/off alternative operation

The final part of the completed circuit will need to be designed to best suit the control needed. The most common output arrangement is the remote on/off alternate action (Figure 8.2). A 4013B flip-flop is used for the latching operation. A capacitor/resistor network on the R input resets the Q output low when power is applied. When the decoder output switches low there is no effect until it rises again, this clocks through the data on the Not Q output which would normally be high. The next decoder pulse will clock through a logic zero. This circuit will 'arm' the alarm when power is applied because the Not Q pin will be high, switching on the BC547 thus pulling down any supply to the ignition control input. Please note the change of resistor value on the alarm ignition input circuit.

Remote on, ignition off (reset)

This method will give better security cover. The alarm will set itself as before but the remote control is needed for disarming the system. The flip-flop is connected as a 'set/reset' bistable. When power is applied to the ignition input the reset pin is pulled high for an instant due to the 1 μF capacitor charging up via the 33 K resistor. This resets the Q output low, turning off the BC547 transistor. The high output on the collector allows the application of a 12 V supply at the alarm ignition terminal to remain high until a pulse is received from the decoder. A decoder pulse received on the set terminal causes the Q output to switch high which will turn off the supply to the alarm ignition control.

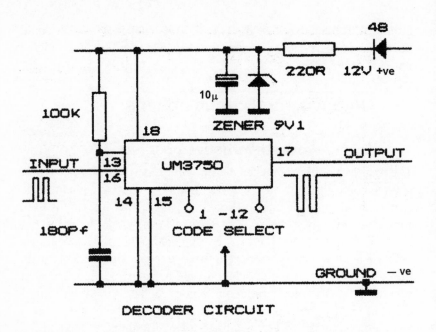

DECODER CIRCUIT

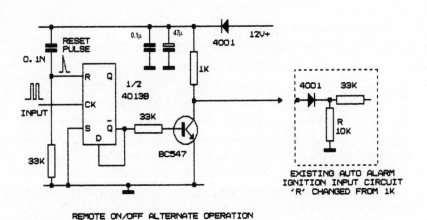

REMOTE ON/OFF ALTERNATE OPERATION

Figure 8.2 Remote control output (circuits) *Source:* GEC Plessey Semiconductors

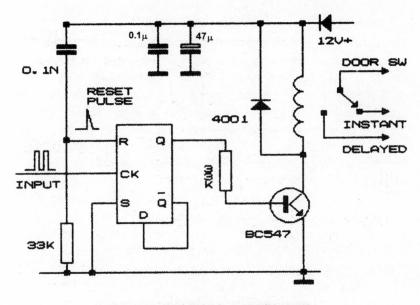

REMOTE ON/OFF ALTERNATE
RELAY DRIVER

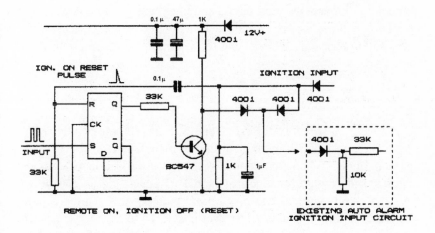

REMOTE ON, IGNITION OFF (RESET)

EXISTING AUTO ALARM
IGNITION INPUT CIRCUIT

Figure 8.2 (cont) Remote control output (circuits)

Remote on/off alternate relay option

This circuit is very similar to the first alternate action remote unit. The main difference is that a relay is used for switching the output. This means that heavy loads can be controlled, i.e. the motor on your garage doors, etc. But for full security on your car the relay can switch the door contacts from an instant to a delayed alarm operation. To do this, instead of using the usual delayed white wire at the switch by itself, take the doorswitch wire to the common relay connection. The white wire would then go to the 'normally open' contact leaving one of the instant wires to be connected to the 'normally closed' contact. Now if the car door is opened without the remote being used the alarm will sound immediately. If the remote had been operated then the normal delayed entry would apply. Now should your car keys were to be lost or stolen then the system would not be an easy one to break. Without knowing how the system works a thief (if he or she found the car) could operate the remote and assume he or she had disabled the alarm. A surprise would then be in store for them!

Printed circuit boards will have to be designed to fit the design used and the enclosures available. This can be most awkward when searching for a suitable transmitter box!

No doubt other uses can be found for these circuits. A little careful thought will reveal the most useful and imaginative ideas. The coding applied to the encoder i.c. could be changed using CMOS logic or a diode matrix, etc., whichever is the most suitable for the application in mind.

Conclusion

Any system which is different from the more common commercial types or even those fitted to a car during manufacture will confuse a potential thief. When common alarms are in use the system becomes familiar and ways are found to beat them. This reduces their effectiveness and makes any car so fitted much more vulnerable. Even the simple pushbutton immobilizer is a very useful addition to any existing alarm system. Take your pick of the various units described and fit more than one to increase your security level.

Appendix

PCBs

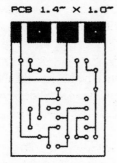

Figure 1.2 Flashing beacon PCB

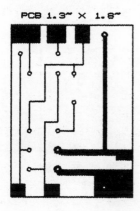

Figure 1.4 Basic pushbutton immobilizer, with optional 33R resistor PCB

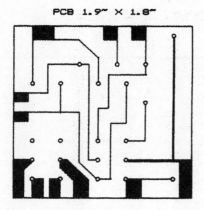

Figure 1.6 Pushbutton immobilizer with full control and siren PCB

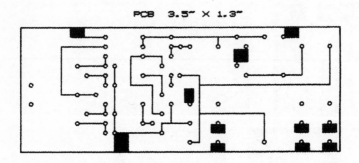

Figure 1.8 Electronic 'key' immobilizer PCB

PCB 3.0" X 3.0"

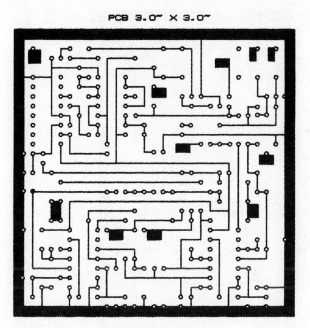

Figure 1.10 Four digit keypad immobilizer PCB

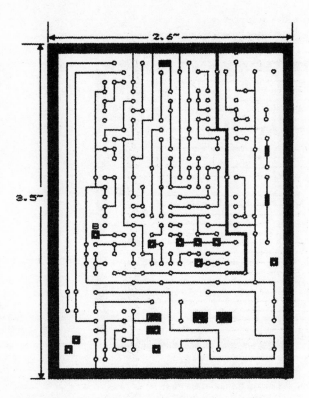

Figure 2.4 Basic alarm system PCB

PCB 4.4" X 3.0"

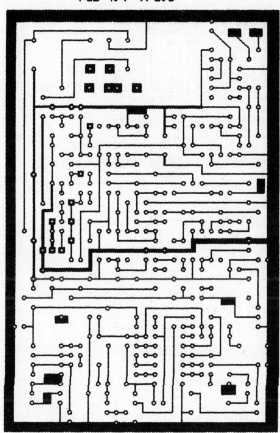

Figure 5.1 Full special alarm system PCB

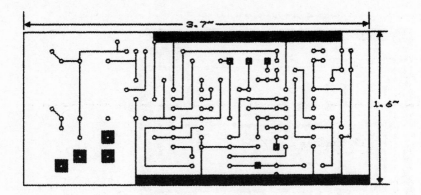

Figure 7.4 Garage alarm PCB

Index